Probability models for data

Unit Guide

The School Mathematics Project

CAMBRIDGE
UNIVERSITY PRESS

Main authors	Chris Belsom
	Robert Black
	David Cundy
	Chris Little
	Jane Southern
	Fiona McGill
Team leader	Chris Belsom
Project director	Stan Dolan

The authors would like to give special thanks to Ann White for her help in producing the trial edition and in preparing this book for publication.

Published by the Press Syndicate of the University of Cambridge
The Pitt Building, Trumpington Street, Cambridge CB2 1RP
40 West 20th Street, New York, NY 10011–4211, USA
10 Stamford Road, Oakleigh, Victoria 3166, Australia

© Cambridge University Press 1992

First published 1992

Produced by Gecko Limited, Bicester, Oxon.

Cover design by Iguana Creative Design

Printed in Great Britain at the University Press, Cambridge

British Library cataloguing in publication data

A catalogue record for this book is available from the British Library.

ISBN 0 521 40880 6

Contents

Introduction to 16–19 Mathematics

Nobody reads introductions and nobody reads teachers' guides, so what chance does the introduction to this Unit Guide have? The least we can do is to keep it short! We hope that you will find the discussion point and tasksheet commentaries and ideas on presentation and enrichment useful.

The School Mathematics Project was founded in 1961 with the purpose of improving the teaching of mathematics in schools by the provision of new course materials. SMP authors are experienced teachers and each new venture is tested by schools in a draft version before publication. Work on *16–19 Mathematics* started in 1986 and the pilot of the course has been used by over 30 schools since 1987.

Since its inception the SMP has always offered an 'after sales service' for teachers using its materials. If you have any comments on *16–19 Mathematics*, or would like advice on its use, please write to:

16–19 Mathematics
The SMP Office
The University
Southampton SO9 5NH

Why 16–19 Mathematics?

A major problem in mathematics education is how to enable ordinary mortals to comprehend in a few years concepts which geniuses have taken centuries to develop. In theory, our view of how to pass on this body of knowledge effectively and pleasurably has changed considerably; but no great revolution in practice has been seen in sixth-form classrooms generally. We hope that in this course, the change in approach to mathematics teaching embodied in GCSE schemes will be carried forward. The principles applied in the course are appropriate to this aim.

- Students are actively involved in developing mathematical ideas.
- Premature abstraction and over-reliance on algorithms are avoided.
- Wherever possible, problems arise from, or at least relate to, everyday life.
- Appropriate use is made of modern technology such as graphic calculators and microcomputers.
- Misunderstandings are confronted and acted upon.
  By applying these principles and presenting material in an attractive way, A level mathematics is made more accessible to students and more meaningful to them as individuals. The *16–19 Mathematics* course is flexible enough to provide for the whole range of students who obtain at least a grade C at GCSE.

Structure of the courses

The A and AS level courses have a core-plus-options structure. Details of the full range of possibilities, including A and AS level *Further Mathematics* courses, may be obtained from the Joint Matriculation Board, Manchester M15 6EU.

For the A level course *Mathematics* (*Pure with Applications*), students must study eight core units and a further two optional units. The structure diagram below shows how the units are related to each other. Other optional units are being developed to give students an opportunity to study aspects of mathematics which are appropriate to their personal interests and enthusiasms.

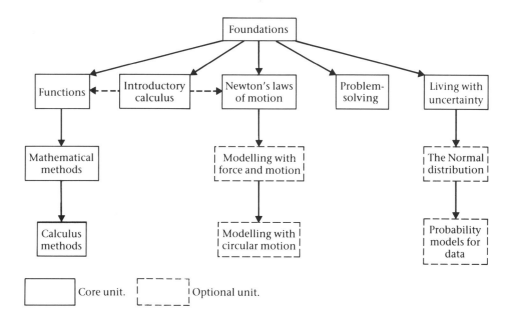

The *Foundations* unit should be started before or at the same time as any other core unit.

Any of the other units can be started at the same time as the *Foundations* unit. The second half of *Functions* requires prior coverage of *Introductory calculus*. *Newton's laws of motion* requires calculus notation which is covered in the initial chapters of *Introductory calculus*.

For the AS level *Mathematics* (*Pure with Applications*) course, students must study *Foundations*, *Introductory calculus* and *Functions*. Students must then study a further two applied units.

Material

The textbooks contain several new devices to aid an active style of learning.

- Topics are opened up through **group discussion points**, signalled in the text by the symbol

and enclosed in rectangular frames. These consist of pertinent questions to be discussed by students, with guidance and help from the teacher. Commentaries for discussion points are included in this unit guide.

- The text is also punctuated by **thinking points**, having the shape

and again containing questions. These should be dealt with by students without the aid of the teacher. In facing up to the challenge offered by the thinking points it is intended that students will achieve a deeper insight and understanding. A solution within the text confirms or modifies the student's response to each thinking point.

- At appropriate points in the text, students are referred to **tasksheets** which are placed at the end of the relevant chapter. A tasksheet usually consists of a self-contained piece of work which is used to investigate a concept prior to any formal exposition. In many cases, it takes up an idea raised in a discussion point, examining it in more detail and preparing the way for formal treatment. There are also **extension tasksheets** (labelled by an E), for higher attaining students, which investigate a topic in more depth and **supplementary tasksheets** (labelled by an S) which are intended to help students with a relatively weak background in a particular topic. Commentaries for all the tasksheets are included in this unit guide.

The aim of the **exercises** is to check full understanding of principles and give the student confidence through reinforcement of his or her understanding.

Graphic calculators/microcomputers are used throughout the course. In particular, much use is made of graph plotters. The use of videos and equipment for practical work is also recommended.

As well as the textbooks and unit guides, there is a *Teacher's resource file*. This file contains: review sheets which may be used for homework or tests; datasheets; technology datasheets which give help with using particular calculators or pieces of software.

Introduction to the unit (for the teacher)

This unit of the *16–19 Mathematics* course is about the modelling of random variables with a number of standard probability models, including the binomial, Poisson and geometric models for discrete data, and the Normal probability model for continuous data. Practical and exploratory work is embodied within the general development of the material, as is relevant computer work. A disc of simple programs to accompany the unit is also available and its use is written in to the textual material.

Chapter 1
Students are encouraged to think about how to measure whether a model is a good fit to the data and build up for themselves the common measure based on observed and expected frequencies. The statistic X^2 is defined and its behaviour investigated in a variety of situations.

Chapter 2
The sampling distribution of X^2 is investigated and the chi-squared distribution defined. The concept of degrees of freedom is encountered in determining which chi-squared distribution is appropriate. The chi-squared test is used in contingency tables.

Chapter 3
Probability models for discrete data are considered, with the geometric, Poisson and binomial models being considered in some detail.

Chapter 4
This chapter is concerned with using the discrete models from chapter 3 as models for data. The use of the χ^2 test to measure the goodness of the fit of the model to the data is also tackled.

Chapter 5
New variables are formed from linear combinations of discrete random variables. The special case of the addition of Poisson variables is analysed. The concept of expectation is introduced formally for the first time and provides a convenient notation. A simple proof for the mean and the variance of the binomial distribution is developed out of ideas considered earlier.

Chapter 6
This chapter considers models for continuous data and formalises the concept of the probability density function. The chapter ends with the extension of the ideas of chapter 5 into the area of continuous data by considering the combination of Normal variables and the variety of practical situations which this encompasses.

Tasksheets and resources

1 Models for data

1.1 Deterministic and probabilistic models

> Which of the following situations are more likely to be successfully modelled with a deterministic mathematical model and which with a probabilistic mathematical model?
>
> (a) the number of matches in a matchbox
>
> (b) the amount the pound in your pocket will be worth in five years' time
>
> (c) the flight of a space probe to the planet Mars
>
> (d) the result of the next general election
>
> (e) the rate of a nuclear reaction
>
> (f) whether a new baby is a boy or a girl

(a) The number of matches in a matchbox would be modelled with a probabilistic mathematical model.

(b) There is a great deal of uncertainty concerning the value of the pound and models to predict its value in the future need to assign probabilities to various possible outcomes. Models are therefore likely to be probabilistic, but once certain assumptions have been made, deterministic models may be used.

(c) The flight of a space probe to the planet Mars can be modelled with a deterministic mathematical model.

(d) The result of the next general election can be modelled with a probabilistic mathematical model.

(e) The rate of a nuclear reaction can be modelled with a deterministic mathematical model as it involves huge numbers of molecules and statistical fluctuations would 'even out'. The underlying process is again probabilistic.

(f) Whether a new baby is a boy or a girl can be modelled with a probabilistic mathematical model.

1.2 Observed and expected frequencies

(a) Are babies equally likely to be male or female?

Do you think that the sex of a first child affects that of a second? How could you find out?

(b) The following table of data concerns 100 families.

	Boy first	Girl first
Boy second	31	21
Girl second	22	26

Do these data affect your answers to (a)?

(a) The likelihood of a baby being a boy is almost the same as its being a girl. If you consider the sex of a child as being independent of the sex of other children in a family, then the sex of the first child would not affect that of subsequent children. You could investigate this assumption by looking at a large number of families and counting the number of male and female children in each family.

(b) These data would seem to suggest that if the first child is a boy then the second child is more likely to be a boy also. It will also seem that the second child is more likely to be a girl if the first child is a girl. In a sample of 100 families, you would expect about 25 in each of the cells. You should consider whether the variation from 25 is sufficient for you to reconsider your ideas in (a). Are you just seeing **random** fluctuations, or do the data indicate something significant?

Random numbers

1 (a) and (b) You may find that your sequence contains more (0, 1) and (1, 0) pairings than (0, 0) and (1, 1).

   (c) If the sequence were random, the probability of obtaining each of the number pairs (0, 0), (0, 1), (1, 0) and (1, 1) would be 0.25. The expected frequencies would be 25 for each pair.

   (d) A possible result might be:

	Expected frequencies	Observed frequencies
(0, 0)	25	16
(0, 1)	25	30
(1, 0)	25	30
(1, 1)	25	24

The sequence obtained naturally produces different results from the expected frequencies. Here, there are fewer (0, 0) pairs than you might expect. Again, this may be acceptable random variation or a conscious effort on your part to avoid following a zero with a zero. It is difficult to decide without some objective measure.

2 (a) Some pairs obtained from a typical calculator-generated random sequence are:

Number pair	Frequency
(0, 0)	29
(0, 1)	24
(1, 0)	25
(1, 1)	22

Obviously, your results will be different.

  (b) (i) and (ii)
In order to decide whether the calculator was producing truly random sequences, you would need to consider some **measure** of the difference between the values you would **expect** if it were fair and those you actually achieve in practice. You will further investigate this **difference measure** in this chapter and in chapter 2.

Measuring bias

1 The probability of throwing a one is $\frac{1}{6}$ so the expected number of ones is $\frac{1}{6} \times 1200 = 200$. Similarly, the expected number of sixes is also 200.

2 Dice A and C are possibly biased, but you cannot be sure. The results for die B seem too good to be true! Die D appears biased because the 'other' scores appear too low.

3 (a)

	1	6	Other
Observed	182	238	780
Expected	200	200	800
Deviation between O and E	−18	38	−20

Sum of deviations = 0

(b) This is not satisfactory as the positive and negative values cancel one another out.

4 (a) Die A: $\Sigma (O - E)^2 = 18^2 + 38^2 + 20^2 = 2168$

Die B: $\Sigma(O - E)^2 = 1^2 + 1^2 + 0^2 = 2$

Die C: $\Sigma(O - E)^2 = 20^2 + 18^2 + 38^2 = 2168$

Die D: $\Sigma(O - E)^2 = 20^2 + 18^2 + 38^2 = 2168$

(b) Comparing results for die C and die D gives the same value, but this calculation does not take into account the fact that die D was only thrown 600 times, whereas die C was thrown 1200 times.

(c) Comparing results for dice A and C also gives the same value, whereas, intuitively, you would expect different results.

5 (a) Die B: $X^2 = \dfrac{(201 - 200)^2}{200} + \dfrac{(199 - 200)^2}{200} + \dfrac{(800 - 800)^2}{800} = 0.01$

Die C: $X^2 = \dfrac{(220 - 200)^2}{200} + \dfrac{(218 - 200)^2}{200} + \dfrac{(762 - 800)^2}{800} = 5.425$

Die D: $X^2 = \dfrac{(120 - 100)^2}{100} + \dfrac{(118 - 100)^2}{100} + \dfrac{(362 - 400)^2}{400} = 10.85$

(b) In order of increasing value of X^2, the dice are B, C, A and D.

2 The chi-squared test

2.1 The distribution of X^2

> (a) How could you use the program to decide if a die is unbiased?
>
> (b) Use the program to help you decide how likely it is that die E is biased.

(a) As a basis for making an 'absolute' judgement on a given die against a fair die, X^2 could be calculated for 800 throws of an unbiased die. Repeating the experiment (i.e. throwing the die a further 800 times) is likely to produce a different value of X^2. This could be simulated on a computer, producing a distribution for X^2.

From the distribution, you could observe the likelihood of obtaining the single value of X^2 which you have for each die.

(b) The program *Chi-squared* enables you to assess how likely a value for X^2 is, on the assumption that A is unbiased. The program calculates the value of X^2, the difference measure, a large number of times. Using *Chi-squared* for 600 throws, you should obtain a value of X^2 as high as 10 only about once every 100 samples. There is a strong likelihood that E is biased.

2.4 Contingency tables

The data suggest that in the 26–40 age group, a larger proportion of people will vote Labour than in the other age groups.

If there were no connection between age and voting intention then you would expect the same proportion of each age group to vote Labour. Hence data here suggest $\frac{66}{218}$ of those questioned will vote for the Labour councillor. Since there are 62 in the age group 26–40 years you would expect:

- $\frac{66}{218} \times 62 = 18.8$ to vote Labour;
- the remaining 43.2 to vote Conservative.

Figures for each of the other cells could be calculated in a similar way.

Counting constraints

1 (a) Total number of girls $= (1 \times 17) + (2 \times 21) + (3 \times 4) = 71$

  (b) In 51 families which contain 3 children there will be a total of 153 children. The proportion of girls is therefore $\frac{71}{153} = 0.464$ (to 3 s.f.).

  (c)

Number of girls per family	0	1	2	3
Probability	0.154	0.400	0.346	0.100
Expected frequency	7.85	20.4	17.65	5.1

2 (a) $X^2 = \dfrac{(9 - 7.85)^2}{7.85} + \dfrac{(17 - 20.4)^2}{20.4} + \dfrac{(21 - 17.65)^2}{17.65} + \dfrac{(4 - 5.1)^2}{5.1}$

   $= 1.61$

  (b) The lower value of X^2 suggests that this second model is a better fit than the binomial model with $p = 0.5$.

3

Number of girls	0	1	2	3
Observed	9	17	x	y

$$9 + 17 + x + y = 51$$
$$x + y = 25$$
$$(0 \times 9) + (1 \times 17) + (2 \times x) + (3 \times y) = 71$$
$$2x + 3y = 54$$

Solving for x and y gives $x = 21$ and $y = 4$.

Voting and age

1

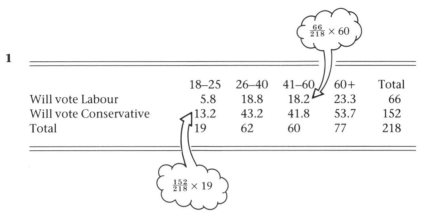

	18–25	26–40	41–60	60+	Total
Will vote Labour	5.8	18.8	18.2	23.3	66
Will vote Conservative	13.2	43.2	41.8	53.7	152
Total	19	62	60	77	218

Cloud annotations: $\frac{66}{218} \times 60$ and $\frac{152}{218} \times 19$

2 $X^2 = 7.75$

3 There are 3 degrees of freedom.
From $\chi^2(3)$ tables:
$\chi^2(3)$ exceeds 6.25 for 10% of samples and it exceeds 7.81 for 5% of samples.
A value of 7.75 is therefore not significant.
There is no evidence in this sample to suggest that voting intention is related to age.

3 Probability distributions for counting cases

3.2 The geometric distribution

> Why is the binomial distribution not a suitable model for the distribution of X?

X has a number of properties in common with a binomial variable. It is a discrete variable, there is a constant probability of the event occuring at each trial and there is independence between trials. However, there is not a **fixed** number of trials (the n in the binomial distribution).

3.3 The binomial distribution revisited

(a) The random variable (X) is the number of boys in the ten cots. X is a discrete variable which may take any of the values 0, 1, 2, . . ., 10. On the assumption that babies are equally likely to be male or female, the appropriate binomial model for X is $X \sim B(10, \frac{1}{2})$.

(b) As discussed in section 3.2, the binomial and geometric variables have a number of factors in common. However, the geometric has an infinite number of possible outcomes, while for a binomial variable, the number of outcomes is finite. The geometric distribution counts the number of cases before an event occurs.

3.4 The Poisson distribution

> What values can X take? Could the distribution of X be binomial, or geometric? What sort of shape do you think the graph of the distribution will take?

Theoretically, X can take any whole number value, although large numbers will be very unlikely. The distribution could not be binomial as it has the possibility of infinitely many outcomes. The distribution is not geometric as it is not waiting for an event to happen.

The graph should have a peak around 4, with events around this being most likely.

It's a girl

1

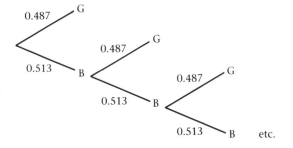

You must assume that the probabilities remain constant throughout – that each birth is independent of all other births.

2 (a) 0.487

(b) $(0.513)(0.487) = 0.250$

(c) $(0.513)^4(0.487) = 0.0337$

(d) $(0.513)^{n-1}(0.487)$

3

x	1	2	3	4	5	6	7	8	...
$P(X = x)$	0.487	0.250	0.128	0.0657	0.0337	0.0173	0.00888		...

4 $P(X \geq 8) = 1 - P(x < 8)$
$= 1 - [P(X = 1) + P(X = 2) + \ldots + P(X = 7)]$
$= 0.00935$

Alternatively, if eight or more children are needed for the first girl, then the first seven must have been boys.

$P(7 \text{ boys}) = (0.513)^7 = 0.00935$

5 (a) If $X = 1$ then the event occurs at the first trial.
So $P(X = 1) = p$

(b) The event of interest occurs on the third trial.
$P(X = 3) = q \times q \times p = q^2p$

(c) $P(X = n) = \boxed{q \times q \times q \times \ldots} \times p = q^{n-1}p$
$n - 1$ non-occurrences

Programming the binomial

1 $\dfrac{r!}{(r-1)!} = \dfrac{r(r-1)(r-2)\ldots 3 \times 2 \times 1}{(r-1)(r-2)\ldots 3 \times 2 \times 1} = r$

2 (a) $\dfrac{(r+1)!}{r!} = r+1$ (b) $\dfrac{(r+1)!}{(r-1)!} = (r+1)r$

   (c) $\dfrac{(r+2)!}{r!} = (r+2)(r+1)$ (d) $\dfrac{(r+n)!}{r!} = (r+n)(r+n-1)\ldots(r+1)$

3 $\dbinom{n}{r+1} \Big/ \dbinom{n}{r} = \dfrac{n!}{(n-r-1)!(r+1)!} \Big/ \dfrac{n!}{(n-r)!r!}$

$$= \dfrac{n!}{(n-r-1)!(r+1)!} \; \dfrac{(n-r)!r!}{n!}$$

$$= \dfrac{(n-r)!}{(n-r-1)!} \; \dfrac{r!}{(r+1)!} = \dfrac{n-r}{r+1}$$

$$\dfrac{P(r+1)}{P(r)} = \dfrac{\dbinom{n}{r+1}p^{r+1}q^{n-r-1}}{\dbinom{n}{r}p^{r}q^{n-r}} = \dfrac{\dbinom{n}{r+1}}{\dbinom{n}{r}}\dfrac{p}{q} = \left(\dfrac{n-r}{r+1}\right)\dfrac{p}{q} \text{ hence the result}$$

Your program should be similar to the one given here.

```
"N"? → N:  "P"? → P
1-P → Q:0 → R
Q xʸ N → X
Lbl 1:R◢
X◢
(N-R) × P × X ÷ ((R+1) × Q) → X
R+1 → R
R≤N ⇒ GOTO 1
```

4 (a)

x	0	1	2	3
$P(X = x)$	0.42	0.42	0.14	0.02

(b)

x	0	1	2	3	4
$P(X = x)$	0.20	0.40	0.30	0.10	0.01

Emergency deliveries

1 (a) The mean of B(n, p) is np. $np = 12 \times \frac{1}{3} = 4$

(b)

x	0	1	2	3	4	5	6	7	8	9	10	11	12
$P(X = x)$	0.008	0.046	0.127	0.212	0.238	0.191	0.111	0.048	0.015	0.003	0.000	0.000	0.000

2 (a) There are 48 intervals of 15 minutes. On average, there are four requests in 12 hours.

$$P(\text{request in a 15-minute interval}) = \frac{4}{48} = \frac{1}{12}$$

(b) $X \sim B(48, \frac{1}{12})$

The distribution of X is as shown. Compare these values with those for B($12, \frac{1}{3}$).

x	0	1	2	3	4	5	6	7	8	9	10	11	12
$P(X = x)$	0.015	0.067	0.143	0.199	0.204	0.163	0.106	0.058	0.027	0.011	0.004	0.000	0.000

3 $X \sim B(720, \frac{1}{180})$

x	0	1	2	3	4	5	6	7	8	9	10	11	12
$P(X = x)$	0.018	0.073	0.146	0.196	0.196	0.156	0.104	0.060	0.030	0.013	0.005	0.000	0.000

4 $X \sim B(7200, \frac{1}{1800})$

x	0	1	2	3	4	5	6	7	8	9	10	11	12
$P(X = x)$	0.018	0.073	0.147	0.196	0.195	0.156	0.104	0.060	0.030	0.013	0.005	0.000	0.000

5 np is the mean value.

For $X \sim B(48, \frac{1}{12})$ the mean is $48 \times \frac{1}{12} = 4$.

Similarly, $np = 4$ for the other distributions.

6 The 4 represents the mean number of requests. In the general expression for $P(X = r)$:

$$P(X = r) = \frac{e^{-\lambda}\lambda^r}{r!} \quad \text{where } \lambda \text{ is the mean value}$$

7 $P(X = 0) = \dfrac{e^{-4}4^0}{0!} = e^{-4} = 0.0183$

$P(X = 1) = \dfrac{e^{-4}4^1}{1} = 4e^{-4} = 0.073$

$P(X = 2) = \dfrac{e^{-4}4^2}{2!} = 0.147$

and so on.

Your results should confirm that the distributions are identical and that the given distribution provides a very good approximation to the binomial under the conditions indicated.

8 $P(X \geqslant 4) = 1 - P(X < 4)$

$\qquad\qquad = 1 - P(X = 0, 1, 2, 3)$

$P(X = 0) = e^{-4} = 0.0183$

$P(X = 1) = 4e^{-4} = 0.073$

$P(X = 2) = \dfrac{4^2 e^{-4}}{2!} = 0.147$

$P(X = 3) = \dfrac{4^3 e^{-4}}{3!} = 0.195$

$P(X \geqslant 4) = 0.566$

From binomial to Poisson

1 $P(X = r + 1) = \dfrac{e^{-\lambda}\lambda^{r+1}}{(r+1)!}$

$\qquad\qquad = \dfrac{\lambda}{(r+1)}\ \dfrac{e^{-\lambda}\lambda^r}{r!} = \dfrac{\lambda}{r+1}\,P(X = r)$

$\qquad P(X = 0) = \dfrac{e^{-\lambda}\lambda^0}{0!} = e^{-\lambda}$

2 $e^{-\lambda} = 1 - \dfrac{\lambda}{1!} + \dfrac{\lambda^2}{2!} - \dfrac{\lambda^3}{3!} + \dots$

3 Define $\lambda = np,\quad$ so $p = \dfrac{\lambda}{n}$.

Now $\quad q = 1 - p,\quad$ so $q^n = \left(1 - \dfrac{\lambda}{n}\right)^n$

$\Rightarrow\ q^n = 1 - \dfrac{n\lambda}{n} + \binom{n}{2}\dfrac{\lambda^2}{n^2} - \binom{n}{3}\dfrac{\lambda^3}{n^3} + \dots$

$\qquad = 1 - \lambda + \dfrac{n(n-1)}{2!}\ \dfrac{\lambda^2}{n^2} - \dfrac{n(n-1)(n-2)}{3!}\ \dfrac{\lambda^3}{n^3} + \dots$

$\qquad = 1 - \lambda + \dfrac{\lambda^2}{2!}\left(1 - \dfrac{1}{n}\right) - \dfrac{\lambda^3}{3!}\left(1 - \dfrac{1}{n}\right)\left(1 - \dfrac{2}{n}\right) + \dots$

4 As $n \to \infty,\ \left(1 - \dfrac{1}{n}\right) \to 1,\ \left(1 - \dfrac{2}{n}\right) \to 1\quad$ and so on.

Then $q^n \to 1 - \lambda + \dfrac{\lambda^2}{2!} - \dfrac{\lambda^3}{3!} + \dots\quad$ i.e. $\quad q^n \to e^{-\lambda}$

5 $\dfrac{(n-r)p}{q} = \dfrac{\left(1 - \dfrac{r}{n}\right)np}{q} = \dfrac{\left(1 - \dfrac{r}{n}\right)\lambda}{q} = \dfrac{\left(1 - \dfrac{r}{n}\right)\lambda}{1 - p} = \dfrac{\left(1 - \dfrac{r}{n}\right)\lambda}{1 - \dfrac{\lambda}{n}}$

$\Rightarrow$ as $n \to \infty,\ \dfrac{(n-r)p}{q} \to \lambda$

so $\quad \dfrac{(n-r)p}{(r+1)q} \to \dfrac{\lambda}{r+1}$

4 Selecting and testing the models

4.1 Choosing a suitable model

> (a) Of the models you have considered for counting cases, which do you think is the most suitable here? Consider carefully the reasons for your choice.
>
> (b) What assumptions need to be made for your chosen model to be suitable?

(a) It is not binomial as there is no fixed value for n, the number of trials. It is not Poisson as $X \neq 0$ and the events do not occur over a continuous interval of time.
The geometric distribution is the most suitable as it is measuring the number of trials before an event occurs.

(b) To use the geometric distribution, you must assume that successive attempts are independent.

Geometric models

1 For 200 trials, you would expect results **similar** to:

1	2	3	4 . . .
20	18	16	14 . . .

2 You must assume that successive numbers are independent. There are 10 possible outcomes. If the numbers are chosen at random then each is equally likely and so P(zero is chosen) $= \frac{1}{10}$.

3 You will find it necessary to group a number of cells. The more data you collect, the better.

4 For your data you should calculate $\dfrac{\text{total number of runs of length one}}{\text{total number of trials}}$.

5 The second model should be a better fit as p is generated by the data.

6 The mean run length of your data should be approximately 10.

Fitting the Poisson model

1 Mean   = 9.4
  Variance = 10.16

2

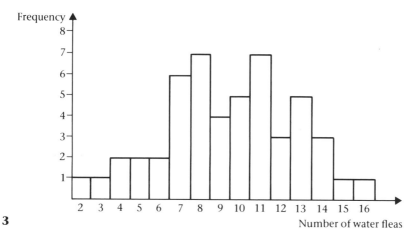

3

Number of water fleas	2	3	4	5	6	7	8	9	10	11	12	13	14	15	16
Probability	0.004	0.011	0.027	0.051	0.079	0.106	0.125	0.131	0.123	0.105	0.082	0.059	0.04	0.025	0.015
Expected	0.18	0.57	1.35	2.53	3.96	5.32	6.25	6.53	6.14	5.25	4.11	2.97	2.00	1.25	0.73

(using $\lambda = 9.4$)

Regroup the data to avoid having expected numbers smaller than 5.

Number of water fleas	2 to 6	7	8	9	10	11	12	13 to 16
Probability	0.172	0.106	0.125	0.131	0.123	0.105	0.082	0.139
Expected	8.59	5.32	6.25	6.53	6.14	5.25	4.11	6.95
Observed	8	6	7	4	5	7	3	10

$$\sum \frac{(O - E)^2}{E} = 3.63$$

There are 8 cells and 2 constraints, so there are 6 degrees of freedom.
$\chi^2(6) = 12.59$ (at the 5% level)
The model appears to be a good fit.

5 Forming new variables

5.1 Games of chance

> The entry price is the same for both games. Which would you choose to play and why?

The mean winnings on the two games will be the same. However, the probability distributions for the two games are very different.

For option 1, the score X can take values from 2 to 12. The probabilities of the extreme scores are:

$$P(X = 2) \;= \tfrac{1}{6} \times \tfrac{1}{6} = \tfrac{1}{36}$$
$$P(X = 12) = \tfrac{1}{36}$$

You could work out the probabilities for each value of X and the distribution would be, approximately:

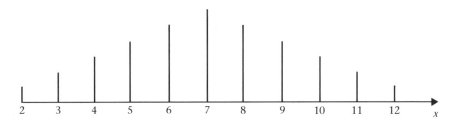

For option 2, all scores are equally likely.

The only possible values of X are 2, 4, 6, 8, 10 and 12 and the distribution is as follows:

You **cannot** alter your expected winnings by your choice of game. Option 2 offers you a greater chance of winning a large amount but, at the same time, a greater chance of winning only a small amount.

5.2 Combining by subtraction

(a) Using the program *DComb*, investigate combinations of variables such as $(R - Y)$, $(B - Y)$, etc. Conjecture a result for the mean and variance of $(X - Y)$ in relation to the mean and variance of X and Y.

(b) List the probability distribution for $(A - B)$ where A and B have probability distributions as follows:

A	0	1	2
P(A)	0.2	0.6	0.2

B	1	2
P(B)	0.5	0.5

(c) Confirm that:

(i) mean$(A - B)$ = mean(A) − mean(B)

(ii) variance$(A - B)$ = variance(A) + variance(B).

(a) Mean $(X - Y)$ = mean (X) − mean (Y)
Variance $(X - Y)$ = variance(X) + variance(Y)

(b) Mean(A) = 1   Mean(B) = 1.5
Variance(A) = 0.4   Variance(B) = 0.25

$A - B$	−2	−1	0	1
P($A - B$)	0.1	0.4	0.4	0.1

Mean$(A - B)$ = −0.5
Variance$(A - B)$ = 0.65

(c) (i) Mean(A) − mean(B) = 1 − 1.5
= −0.5 = mean$(A - B)$

(ii) Variance(A) + variance(B) = 0.4 + 0.25
= 0.65 = variance$(A - B)$

Combining variables

TASKSHEET COMMENTARY 1

1

		Score on blue card 1	2	3
Score on yellow card	1	2	3	4
	2	3	4	5
	3	4	5	6
	4	5	6	7

2 (a) $P(Y + B = 5) = P(Y_4, B_1) + P(Y_3, B_2) + P(Y_2, B_3)$
$= 0.1 \times 0.25 + 0.2 \times 0.5 + 0.3 \times 0.25 = 0.20$
$P(2) = P(Y_1, B_1) = 0.10$
$P(4) = P(Y_2, B_2) + P(Y_3, B_1) + P(Y_1, B_3) = 0.30$
$P(7) = P(Y_4, B_3) = 0.025$

Score $Y + B$	Probability
2	0.10
3	0.275
4	0.30
5	0.20
6	0.10
7	0.025

(b) Mean $(Y + B) = 4.0$ Variance$(Y + B) = 1.5$
Note that mean$(Y + B)$ = mean(Y) + mean(B)
and variance$(Y + B)$ = variance(Y) + variance(B)

3 (a), (b) You should observe that, in each case investigated:

Mean$(X + Y)$ = mean(X) + mean(Y)
Variance$(X + Y)$ = variance(X) + variance(Y)

4 Mean$(aX) = a \times$ mean(X) Variance$(aX) = a^2 \times$ variance(X)

5 (a) Mean$(2R + Y)$ = 2 mean(R) + mean (Y)
Variance$(2R + Y) = 2^2$ variance(R) + variance(Y)

(b) Mean$(3R + 2Y)$ = 3 mean(R) + 2 mean(Y)
Variance$(3R + 2Y) = 3^2$ variance$(R) + 2^2$ variance(Y)

(c) Mean$(2R + 3B + Y)$ = 2 mean(R) + 3 mean(B) + mean(Y)
Variance$(2R + 3B + Y) = 2^2$ variance$(R) + 3^2$ variance(B) + variance(Y)

Some proofs

The geometric distribution

1 6 throws.

2 $(1 - q)^{-2} = 1 + (-2)(-q) + \dfrac{(-2)(-3)}{2!}(-q)^2 + \ldots$

$\qquad\qquad\quad = 1 + 2q + 3q^2 + \ldots$

The Poisson distribution

$E[X] = \sum x_i P(x_i)$ where $P(X = r)\ \dfrac{e^{-\lambda}\lambda^r}{r!}$

$E[X] = \dfrac{0e^{-\lambda}\lambda^0}{0!} + \dfrac{1e^{-\lambda}\lambda}{1!} + \dfrac{2e^{-\lambda}\lambda^2}{2!} + \dfrac{3e^{-\lambda}\lambda^3}{3!} + \ldots$

$\qquad = e^{-\lambda}\left(\lambda + \dfrac{2\lambda^2}{2!} + \dfrac{3\lambda^3}{3!} + \ldots\right)$

$\qquad = \lambda\, e^{-\lambda}\left(1 + \dfrac{\lambda}{1!} + \dfrac{\lambda^2}{2!} + \ldots\right)$

$\qquad = \lambda\, e^{-\lambda}e^{\lambda} = \lambda$

3 Taking the right-hand side of this equation

$E[X(X - 1)] + E[X] = E[X^2 - X] + E[X]$
$\qquad\qquad\qquad\qquad = E[X^2] - E[X] + E[X] = E[X^2]$

4 (a) $E[X(X - 1)] = \displaystyle\sum_{\text{all } x}[x(x - 1)]P(x)$

$\qquad\qquad = 2 \times 1e^{-\lambda}\dfrac{\lambda^2}{2!} + 3 \times 2e^{-\lambda}\dfrac{\lambda^3}{3!} + 4 \times 3e^{-\lambda}\dfrac{\lambda^4}{4!} + \ldots$

$\qquad\qquad = \lambda^2 e^{-\lambda}\left(1 + \dfrac{\lambda}{1!} + \dfrac{\lambda^2}{2!} + \ldots\right)$

$\qquad\qquad = \lambda^2 e^{-\lambda}e^{\lambda} = \lambda^2$

(b) $V[X] = E[X(X - 1)] + E[X] - (E[X])^2$
$\qquad\quad = \lambda^2 + \lambda - \lambda^2 = \lambda$

6 Continuous random variables

6.1 The Normal probability density function

A machine is set to deliver sugar into bags. The weight of sugar it delivers is Normally distributed, having a mean of 1.1 kg and standard deviation 0.1 kg.

(a) **Approximately** what proportion of bags marked 1 kg will be underweight?

(b) Confirm your answer to (a) by calculating this proportion using Normal tables.

It is useful to be able to make approximate calculations with the Normal distribution based on the knowledge that 68% of the values should be within one standard deviation of the mean and 98% within two standard deviations. Recall also that a sketch always helps in solving problems on the Normal distribution.

(a) Underweight bags are at least one standard deviation below the mean. Approximately:

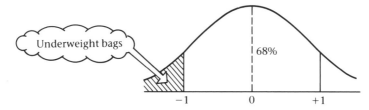

So 32% (approximately) are more than one standard deviation, with 16% beyond +1 and 16% below −1 standard deviation.

About 16% of bags are underweight.

(b)

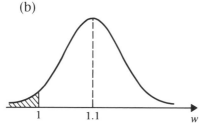

Let W = weight of sugar

$$z = \frac{1 - 1.1}{0.1} = -1$$

$$P(W < 1.0\,\text{kg}) = \Phi(-1)$$
$$= 1 - \Phi(1)$$
$$= 1 - 0.8413$$
$$= 15.9\% \quad (\text{to 3 s.f.})$$

6.4 **The uniform distribution**

(a) If the distance from Penrith to Appleby is given as 13 miles, then what is the probability that the actual distance is:

    (i) between 13 and 13.5 miles;

    (ii) between 13.25 and 13.5 miles?

(b) Describe the probability distribution of X.

(a) (i) The actual distance is as likely to be between 12.5 and 13 as it is to be between 13 and 13.5.

   (ii) 0.25

(b) In general, if M miles is recorded, then the actual distance is between $M - 0.5$ and $M + 0.5$. The error, X, will always be between -0.5 and $+0.5$ miles. No particular range of values of the error is more likely than any other.

The probability of an error on the range 0.1 to 0.2 is the same as for the range 0.2 to 0.3. The probability will therefore be equally distributed across the range.

The probability density function f(x) would be:

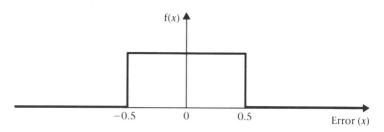

Where probability is equally distributed across a range of values the probability density function is called a **uniform distribution**.

6.5 **The exponential distribution**

A computer bleeps at random once in a second. Having bleeped, how long on average do you expect to wait before the next bleep? Sketch what you think the distribution of T, the length of wait, might look like.

(The program *Bleep 2* will help.)

Bleeps occur randomly.

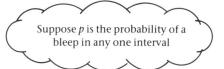

Suppose p is the probability of a bleep in any one interval

The probability of a bleep in the first time interval is p.
The probability of one in the second but not in the first is $(1 - p)p$ because the bleeps occur independently of each other.

Now $p > (1 - p)p$ because $0 < p < 1$ and so the first bleep is more likely to occur in the first interval than in the second.

For example, if $p = 0.01$ then:

P(bleep in 1st interval) = 0.01

and:

P(first bleep in 2nd interval) = $0.99 \times 0.01 = 0.0099$

The first interval is the most likely, the next interval is the second most likely and so on.

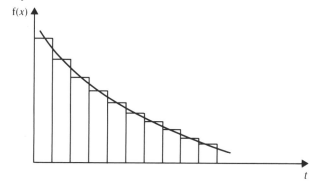

This looks like the graph of an exponential function.

Combining Normal variables

1 (a) $E[X + Y] = E[X] + E[Y]$
$$= 10 + 5 = 15$$

   (b) $V[X + Y] = V[X] + V[Y]$
$$= 1 + 1 = 2$$

2 The distribution should be approximately symmetric about the mean value (15). About 68% of values should be within ± 1 standard deviation and 98% within ± 2 standard deviations of the mean.

3 (a) Your results here will depend on your own computer output. However, you should find that:

   (i) about 50% are less than the mean value;

   (ii) about 68% are between ± 1 standard deviation of the mean;

   (iii) about 98% are between ± 2 standard deviations of the mean;

   (iv) there are almost no values beyond 3 standard deviations of the mean.

   (b) Your results should be close enough to convince you that the distribution is Normal. If you are not convinced, then run the program again with a larger number ($>$ 300) of values. (It will take a while to run!)

4 You should discover that for every combination of the kind $aX \pm bY$, the distribution is Normal when both X and Y are independently distributed Normal variables. For the combinations provided, the results are:

	Variables	Mean	Variance
(a)	$X - Y$	5	2
(b)	$2X$	20	4
(c)	$3X$	30	9
(d)	$2X + Y$	25	5

The resulting distribution is Normal in each case. Your results should (approximately) confirm this.